CONTENTS

CHAPTER 1 • Sorting and Classifying

Explore Sorting 1
Sort by Color, Shape, Size 2
Sort by a Rule 3
Guess My Rule 4
Problem-Solving Strategy: Use Logical Reasoning 5

CHAPTER 2 • Geometry

Explore 3-Dimensional Shapes 6
Use 3-Dimensional Shapes 7
Explore 2-Dimensional Shapes 8
Use 2-Dimensional Shapes 9
Make Shapes 10
Problem-Solving Strategy: Use a Physical Model 11

CHAPTER 3 • Number Readiness and Graphing

Explore Same Number 12
More 13
Fewer 14
Make a Concrete Graph 15
Use a Graph 16
Problem-Solving Strategy: Use a Graph 17

CHAPTER 4 • Numbers to 5

Explore Numbers to 5	18
1 and 2	19
3 and 4	20
5 and 0	21
Numbers 0 to 5	22
Order 1 to 5	23
Problem-Solving Strategy: Make a List	24

CHAPTER 5 • Patterns

Explore Patterns	25
Rhythm Patterns	26
Shape Patterns	27
Number Patterns	28
Problem-Solving Strategy: Use a Pattern	29

CHAPTER 6 • Numbers to 10

Explore Numbers 0 to 10	30
Six	31
Seven	32
Eight	33
Nine	34
Ten	35
Order 1 to 10	36
Problem-Solving Strategy: Make a Pattern	37

CHAPTER 7 • Position and Movement

Explore Inside, Outside, On 38

Top, Middle, Bottom 39

Before, After, Between 40

Left and Right 41

Problem-Solving Strategy: Use a Physical Model 42

CHAPTER 8 • Numbers to 12

Explore Ordinal Numbers 43

Ordinal Numbers 44

Write 0, 1, 2, 3 45

Write 4, 5, 6 46

Write 7, 8, 9, 10 47

Compare Numbers 0 to 10 48

11 and 12 49

Problem-Solving Strategy: Use a Graph 50

CHAPTER 9 • Money and Time

Explore Coins 51

Penny, Nickel, Dime 52

Groups of Coins 53

Order Events 54

More Time, Less Time 55

Clock 56

Problem-Solving Strategy: Use Logical Reasoning 57

CHAPTER 10 • Comparing and Measuring

Explore Size	58
Length	59
Measure Length	60
Weight	61
Capacity	62
Temperature	63
Explore Fair Shares	64
Halves of Regions	65
Problem-Solving Strategy: Guess and Test	66

Chapter 11 • Exploring Adding and Subtracting

Explore Parts and Totals	67
Find Two Parts	68
Find a Total	69
Beginning to Add	70
Adding	71
Beginning to Subtract	72
Subtracting	73
Problem-Solving Strategy: Choose the Operation	74

Chapter 12 • Exploring Greater Numbers

Explore Numbers Greater Than 10	75
Numbers 13 to 15	76
Numbers 16 to 19	77
Numbers to 31	78
Calendar	79
Problem-Solving Strategy: Draw a Picture	80

Name: _____

Practice 1

Explore Sorting

Draw a ring around the objects that are alike. Write an X on the object that is different.

Name:

Practice 2

Sort by Color, Shape, Size

Write an X on objects that are the same shape. Find objects that would be yellow and color them. Find objects that would be green and color them. Draw a line to connect objects that are the same size.

2 • Practice

Grade K, Chapter 1, Lesson 1.2

Name: _____

Sort by a Rule

Practice 3

Follow a rule and sort the pandas into two groups. Color or draw to show the rule.

Grade K, Chapter 1, Lesson 1.3

Name:

Guess My Rule

Practice 4

What is the sorting rule? Draw a ring around the animal that belongs in the group.

Practice — Grade K, Chapter 1, Lesson 1.4

Name:

Problem-Solving Strategy: Use Logical Reasoning

Practice 5

Listen to the clues. Draw a ring around the animals you find.
1. Find the monkey that is sleeping.
2. Find the adult tiger that is drinking.
3. Find the spotted frog that is jumping.
4. Find the striped butterfly that is flying.

Grade K, Chapter 1, Lesson 1.5

Name: _____

Mixed Review

Practice 6

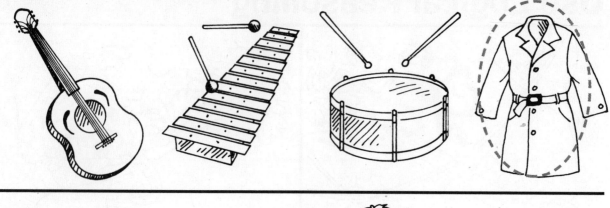

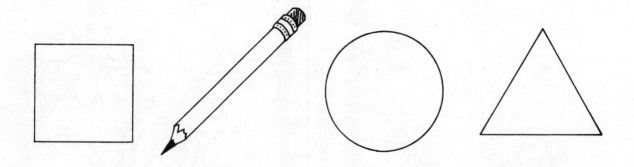

Draw a ring around the object that does not belong.

Grade K, Chapter 2, Lesson 2.1

Name:

Practice 7

Use 3-Dimensional Shapes

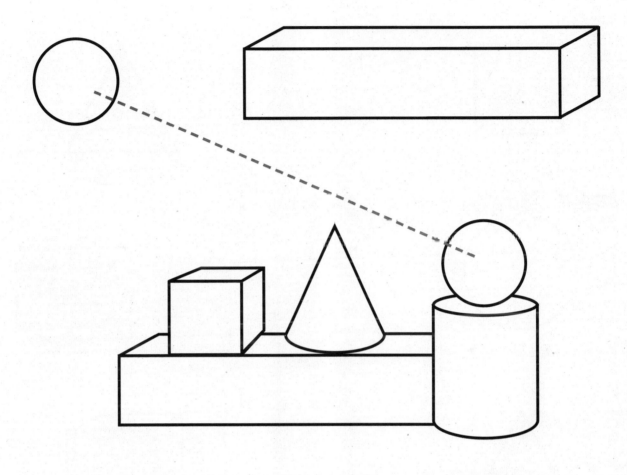

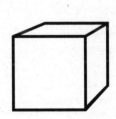

Draw a line from one of the shapes on the outside to the same shape in the structure.

Name:

Explore 2-Dimensional Shapes

Practice 8

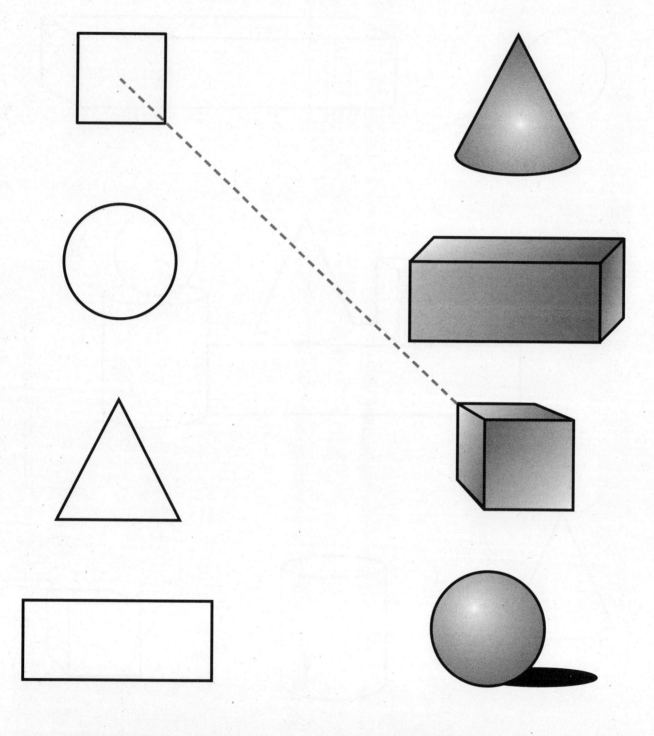

Draw a line to match each two-dimensional shape to a three-dimensional shape.

Grade K, Chapter 2, Lesson 2.3

Name: _____

Use 2-Dimensional Shapes

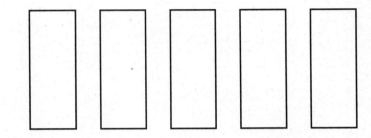

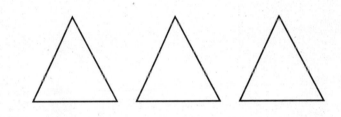

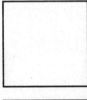
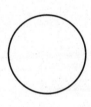
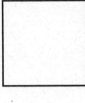
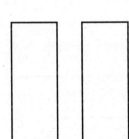
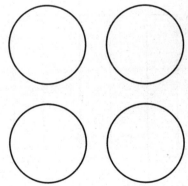

Use the same color crayon to color the matching shapes.

Grade K, Chapter 2, Lesson 2.4

Name: _____

Make Shapes

Practice 10

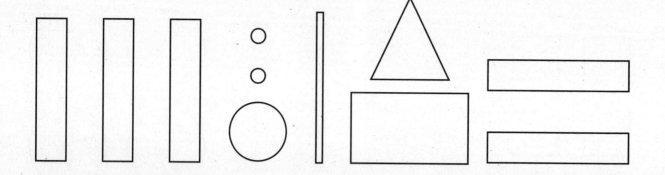

Use each shape only one time to draw a picture. Mark an X on the shape to show you have used it.

Grade K, Chapter 2, Lesson 2.5

Name:

Problem-Solving Strategy:
Use a Physical Model

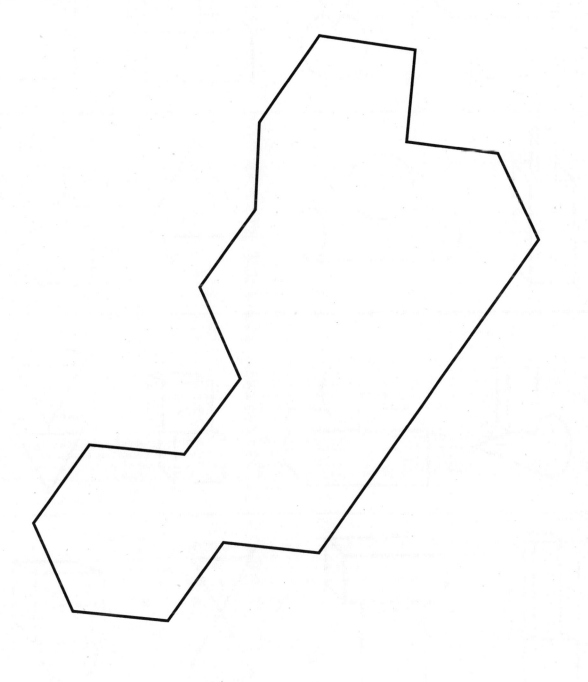

Use pattern blocks to fill in the shape. Color the shapes in your design.

Name:

Mixed Review

Practice 12

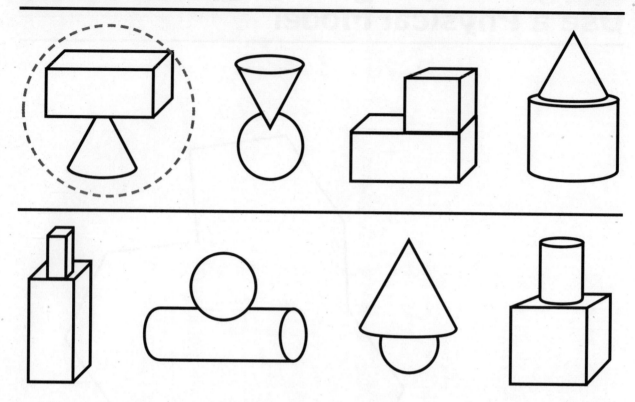

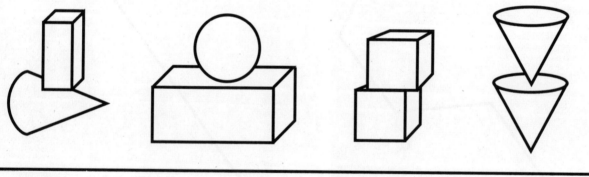

Draw a ring around the shapes that cannot stack.

Grade K, Chapter 3, Lesson 3.1

More

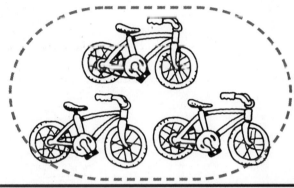

Draw a ring around the group that has more.

Name:

Fewer

Practice 14

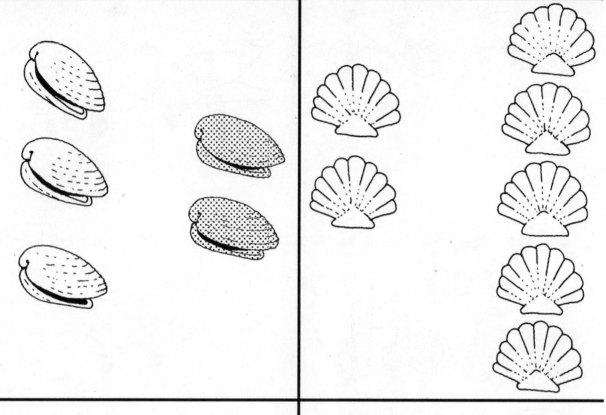
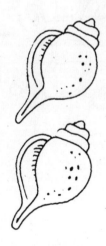

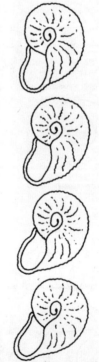

Color the group that has fewer.

Name: _____

Make a Concrete Graph

Practice 15

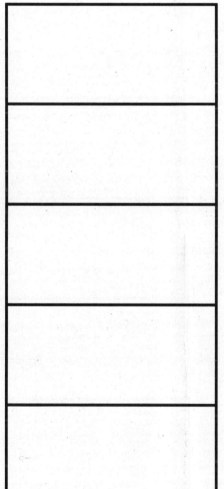
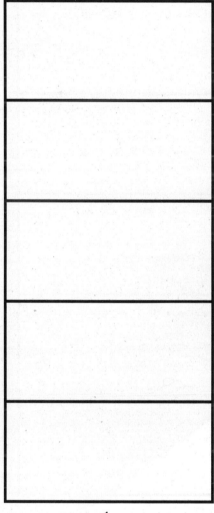

Color some apples red and some apples green. Make a graph to show red and gr___

Grade K, Chapter 3, Lesson 3.4

Name: _____

Use a Graph

Practice 16

 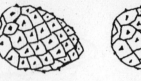 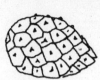

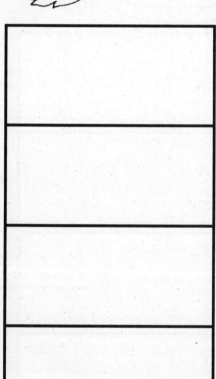
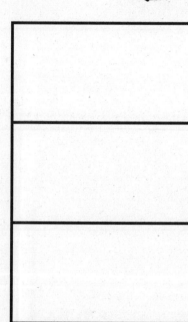

_____ collection has more.

Grade K, Chapter 3, Lesson 3.5

Name: _____

Problem-Solving Strategy: Use a Graph

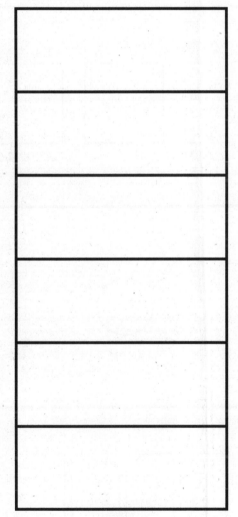
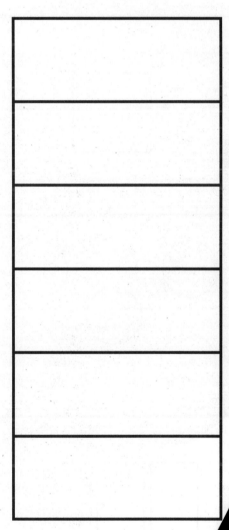

Color squares in graph to find out if there are more striped marbles than plain marbles.

Grade K, Chapter 3, Lesson 3.6

Name:

Mixed Review

Practice 18

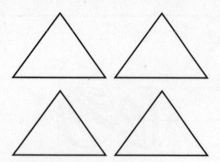 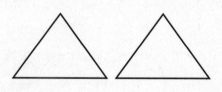

 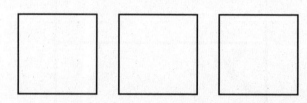

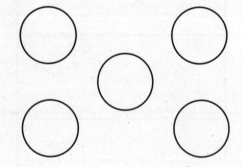 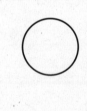

 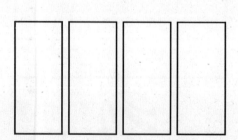

has more. Put an X on the set that has fewer.

Grade K, Chapter 4, Lesson 4.1

Name: _____

1 and 2

1			
1	   		
2	   		
1	   		
2	  		
2	  		

Color to show each number.

Grade K, Chapter 4, Lesson 4.2

Name:

3 and 4

Practice 20

When you see 3 of the same buttons in a row, color them yellow.
When you see 4 of the same buttons in a row, color them green.

Name:

5 and 0

Practice 21

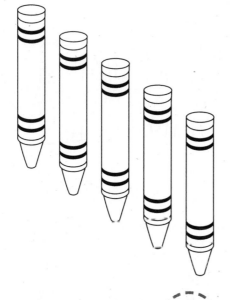

0   (5)

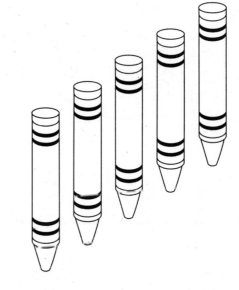

4   5

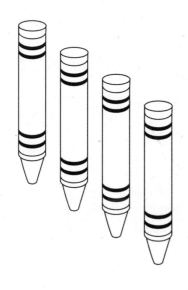

4   5

5   0

Draw a ring around the number of crayons in each set.

Grade K, Chapter 4, Lesson 4.4

Name: _____

Numbers 0 to 5

Practice 22

1 and 2

3 and 2

2 and 2

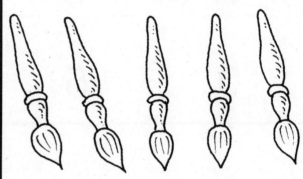

4 and 1

3 and 1

2 and 3

Color the pictures to show the 2 parts.

22 • Practice

Grade K, Chapter 4, Lesson 4.5

Name: _____

Order 1 to 5

Practice 23

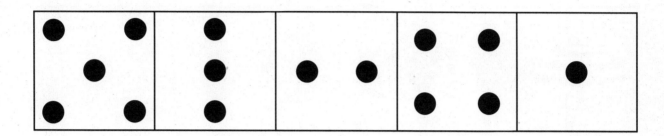

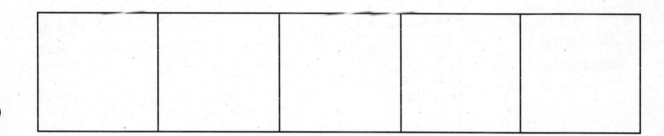

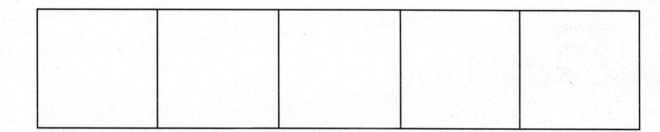

Draw the number cubes in the correct order. Make your own 1 to 5 design.

Grade K, Chapter 4, Lesson 4.6

Problem-Solving Strategy: Make a List

Practice 24

Draw to show all the combinations.

Name: _____

Practice 25

Mixed Review

0 1 (2) 3 4 5

0 1 2 3 4 5

0 1 2 3 4 5

0 1 2 3 4 5

0 1 2 3 4 5

0 1 2 3 4 5

Draw a ring around the number that tells how many fingers you see.

Grade K, Chapter 5, Lesson 5.1 — Practice • 25

Name: _____

Rhythm Patterns

Practice 26

Use 2 colors. Color the cube train to show the pattern of the dancers.

Name: _____

Shape Patterns

Practice 27

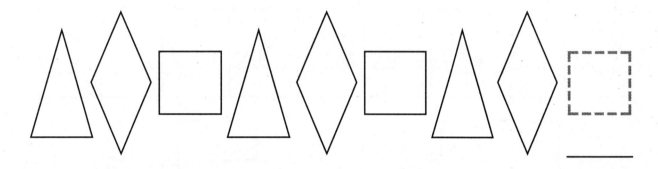

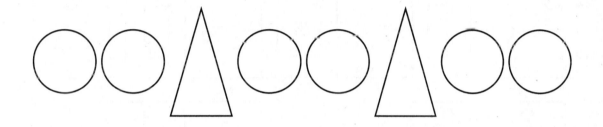

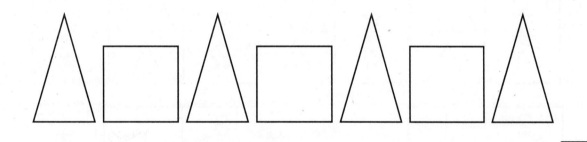

Draw the shape that comes next in the pattern.

Grade K, Chapter 5, Lesson 5.3

Name:

Number Patterns

Practice 28

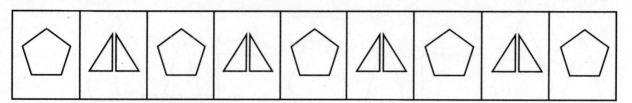

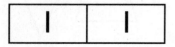 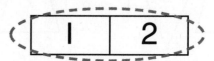

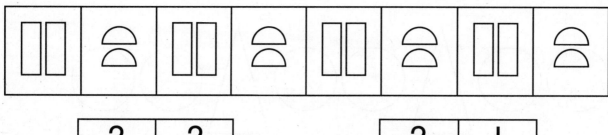

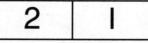

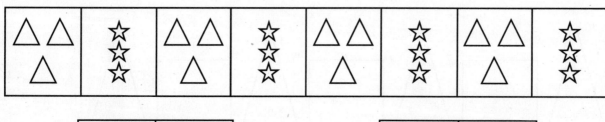

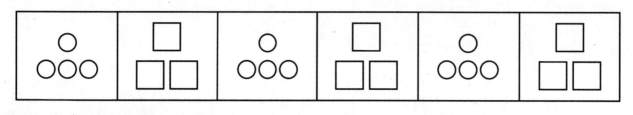

Draw a circle around the number rule for each pattern.

28 • Practice

Grade K, Chapter 5, Lesson 5.4

Name:

Problem-Solving Strategy: Use a Pattern

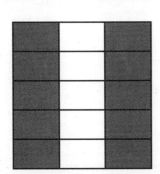

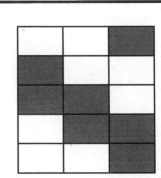

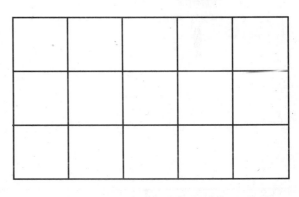

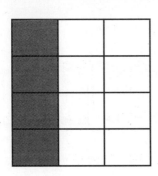

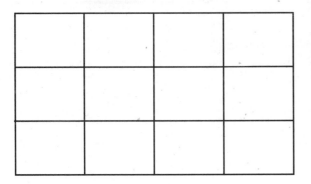

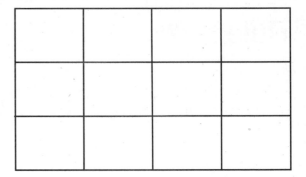

Use colors to transfer the pattern you see to the grid.

Grade K, Chapter 5, Lesson 5.5

Name:

Mixed Review

Practice 30

Rule 1 3

Rule 3 4

Rule 2 6

Rule 4 9

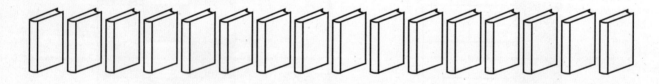

Use 2 colors to show the pattern.

Grade K, Chapter 6, Lesson 6.1

Name: _____

Six

Practice 31

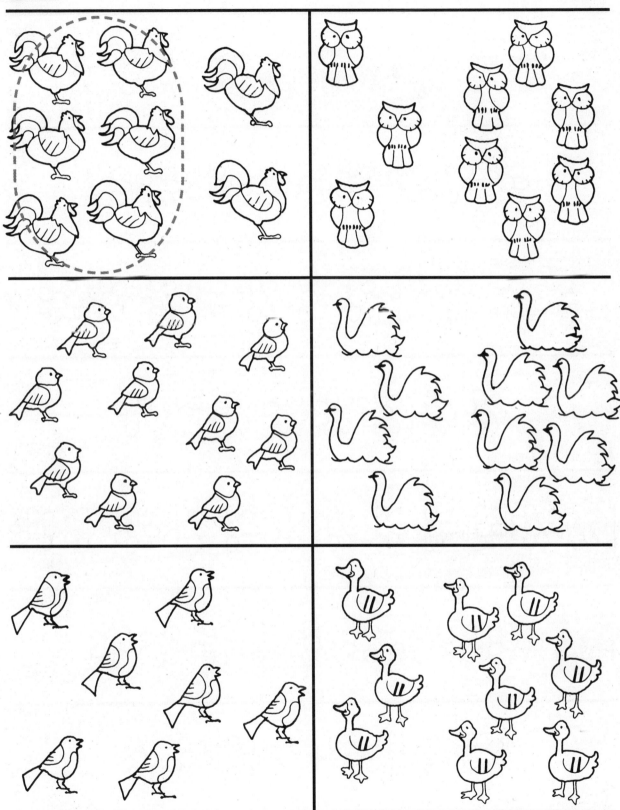

Draw a ring around 6 birds in each group.

Grade K, Chapter 6, Lesson 6.2

Name: _____

Seven

Color 7 in each group.

Name:

Eight

Practice 33

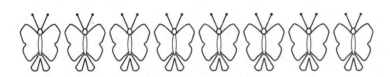   6

   7

   8

   8

   6

   8

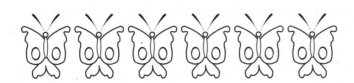   7

Draw lines to match each group with a number.

Grade K, Chapter 6, Lesson 6.4

Name:

Nine

Practice 34

7 8 9

7 8 9

7 8 9

7 8 9

7 8 9

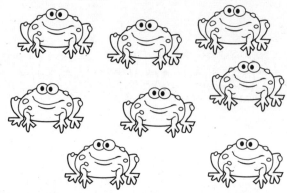
7 8 9

Draw a ring around the number to show how many.

Name:

Ten

Draw more dots to show 10 on each card.

Practice 36

Name: _____

Order 1 to 10

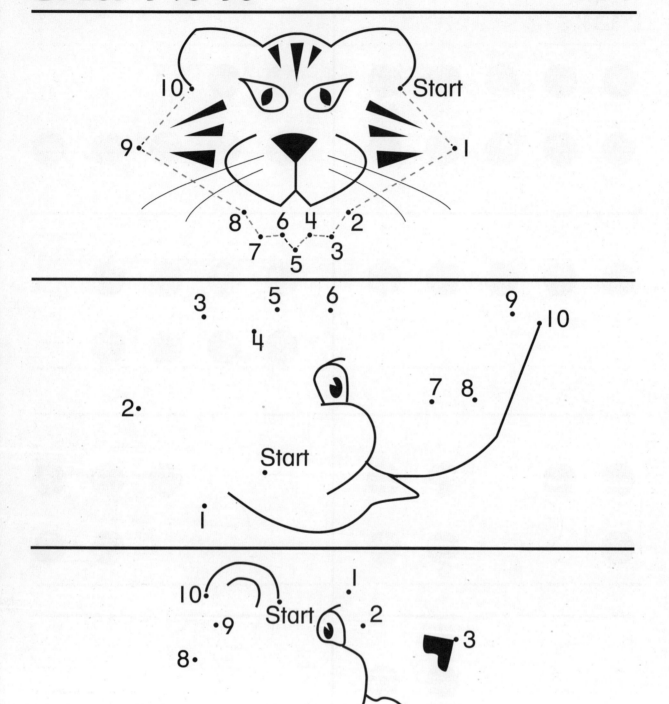

Connect dots from 1 to 10 in order.

36 • Practice Grade K, Chapter 6, Lesson 6.7

Problem-Solving Strategy: Make a Pattern

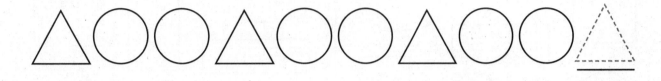

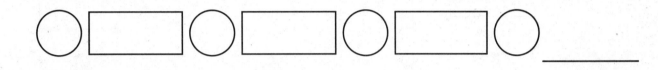

Draw a shape to complete the pattern.
Make a pattern. Use the rule: 2 1 .

Mixed Review

Practice 38

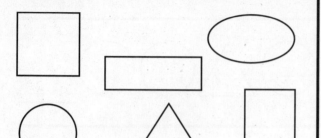
 7 8

6 7 8

7 8 9

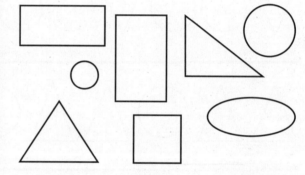
7 8 9

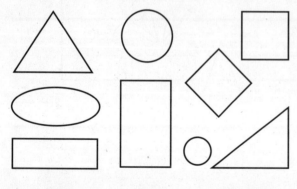
8 9 10

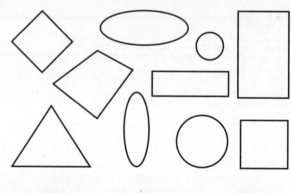
8 9 10

Draw a ring around the number that tells how many.

Grade K, Chapter 7, Lesson 7.1

Name:

Practice 39

Top, Middle, Bottom

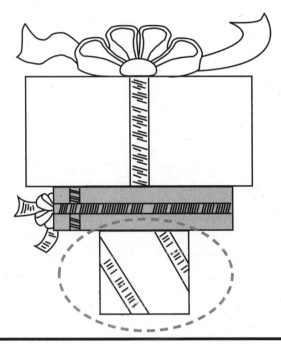
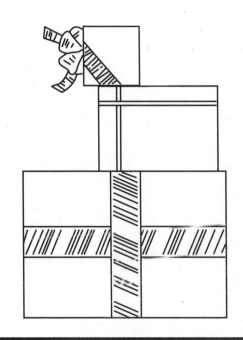
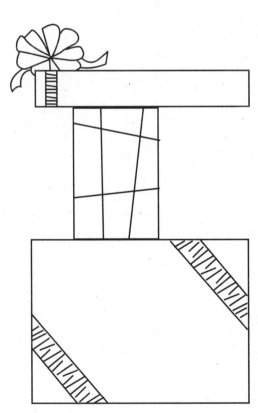
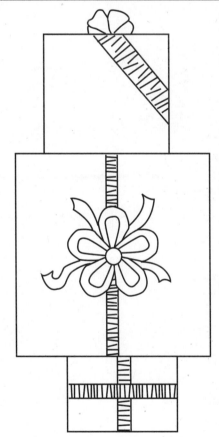

Color the middle box red. Draw a circle around the box at the bottom.

Grade K, Chapter 7, Lesson 7.2

Name:

Before, After, Between

Practice 40

Put an X between 2 cats.
Color the dog that is after the pony.

Grade K, Chapter 7, Lesson 7.3

Name: _____

Left and Right

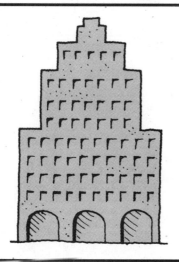

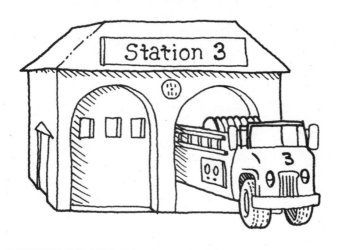

Draw a ring around the building on the right.
Color the building on the left.

Grade K, Chapter 7, Lesson 7.4

Name:

Practice 42

Problem-Solving Strategy: Use a Physical Model

4   5   6

Draw pictures to solve.
3 pandas are in the forest.
2 pandas are at the lake.
How many pandas are there in all?
Draw a ring around the answer.

42 • Practice   Grade K, Chapter 7, Lesson 7.5

Name: _____

Mixed Review

Practice 43

Draw some flowers between the bird bath and the tree. Draw a bird inside the bird bath. Draw grass under the bird bath and the tree. Draw a sun above the bird bath.

Grade K, Chapter 8, Lesson 8.1

Name:

Ordinal Numbers

Practice 44

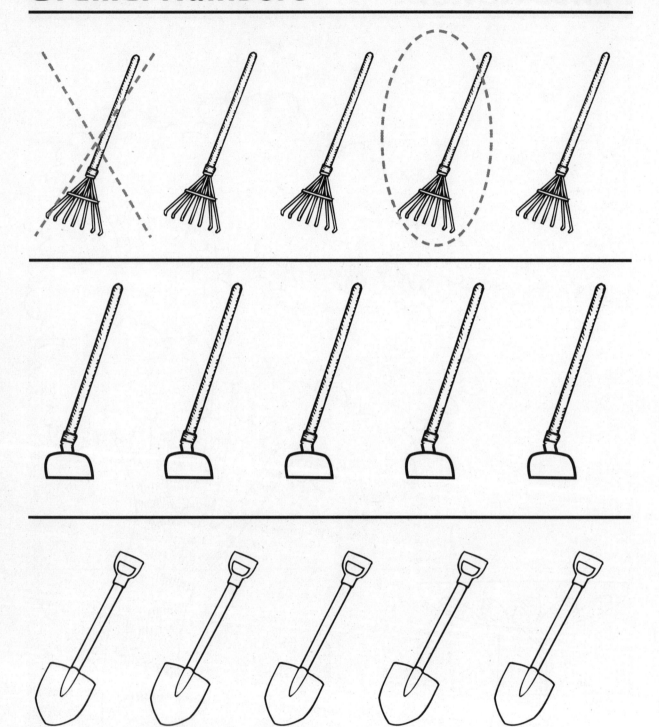

Draw a ring around the fourth and fifth rake.
Draw a ring around the second and third hoe.
Draw a ring around the fifth shovel.
Draw an X on the first object in each row.

Grade K, Chapter 8, Lesson 8.2

Name: _____

Practice 45

Write 0, 1, 2, 3

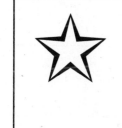   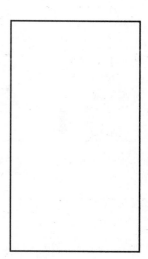

_____   _____   _____   _____

• Trace the number. Draw a line to the correct picture. Then write the correct number under each picture.

Name: _____

Write 4, 5, 6

Practice 46

 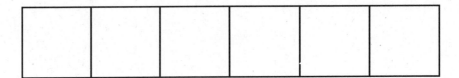

 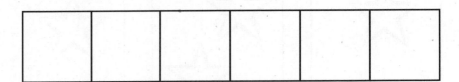

 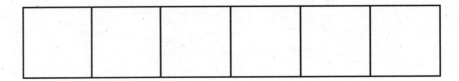

Trace the number. Write the number on the line in front of the boxes. Then color in the boxes to show the number.

46 • Practice

Grade K, Chapter 8, Lesson 8.4

Name: _____

Practice 47

Write 7, 8, 9, 10

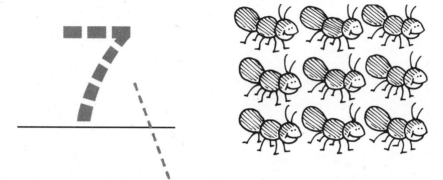

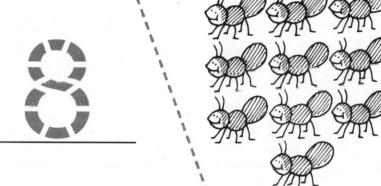

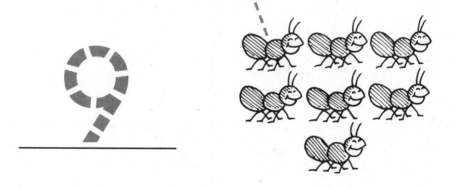

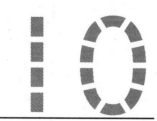 

Trace the numbers and draw a line to the correct set. Write the number next to the set.

Grade K, Chapter 8, Lesson 8.5

Name:

Practice 48

Compare Numbers 0–10

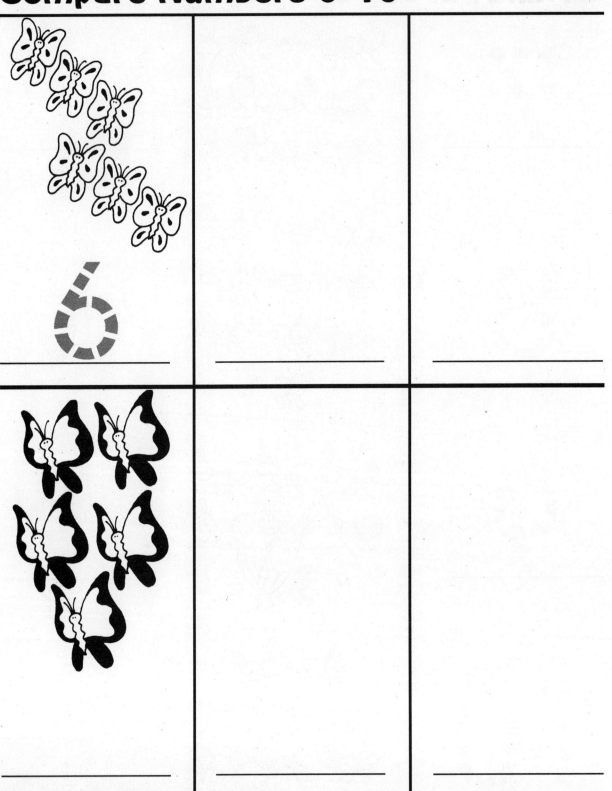

Write the number for the group. Draw a group with fewer butterflies. Write the number. Draw a group with more butterflies. Write the number.

Grade K, Chapter 8, Lesson 8.6

Name: _____

11 and 12

Practice 49

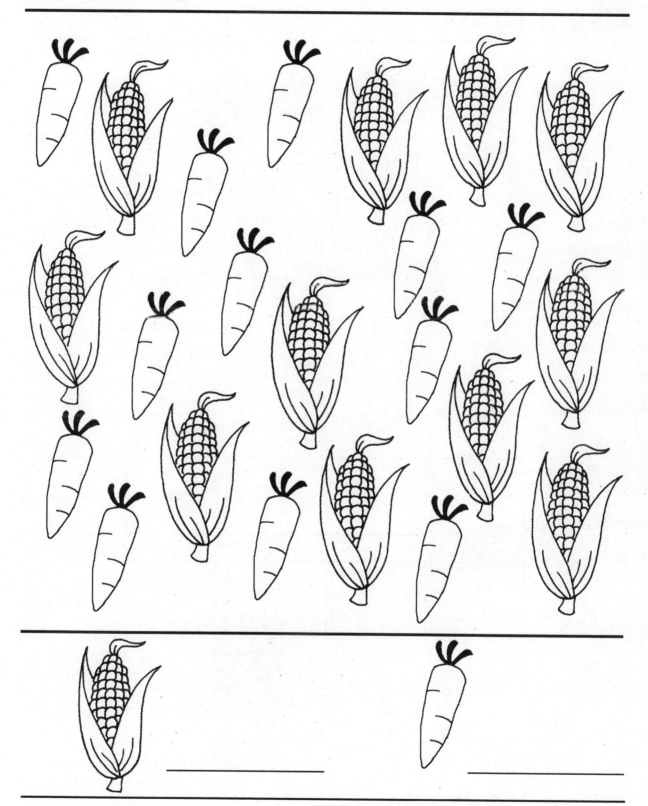

Color each kind of vegetable a different color. Write the number to show how many of each kind of vegetable there is.

Grade K, Chapter 8, Lesson 8.7

Practice • 49

Name:

Practice 50

Problem-Solving Strategy: Use a Graph

_____   _____

Are there more daisies or tulips in the vase? Color to make a graph. Write how many.

50 • Practice

Grade K, Chapter 8, Lesson 8.8

Name:

Practice 51

Mixed Review

Color the fifth can of food. Draw a ring around the eighth jar of juice.
Write the number of cereal boxes in the circle. Write the number of jars of juice in the square.

Grade K, Chapter 9, Lesson 9.1

Name:

Penny, Nickel, Dime

Practice 52

1¢   5¢   10¢

___¢   ___¢   ___¢

___¢   ___¢   ___¢

___¢   ___¢   ___¢

Write the number of cents under each coin.

Practice • Grade K, Chapter 9, Lesson 9.2

Groups of Coins

Practice 53

Name:

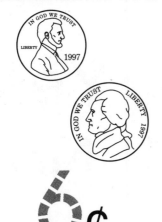
6¢

_____ ¢

_____ ¢

_____ ¢

_____ ¢

_____ ¢

What is each set of coins worth? Write the number of cents on the line.

Grade K, Chapter 9, Lesson 9.3

Name:

Practice 54

Order Events

			Yes No
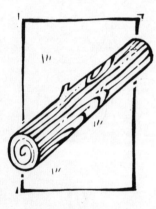			Yes No
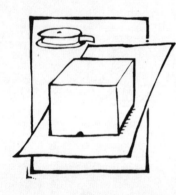			Yes No

Draw a ring around the word **Yes** if the pictures show the events in order. Draw a ring around the word **No** if the pictures are out of order.

Grade K, Chapter 9, Lesson 9.4

Name: _____

More Time, Less Time

Practice 55

Draw a ring around the activity that takes less time to do.

Grade K, Chapter 9, Lesson 9.5

Name:

Clock

Practice 56

1.
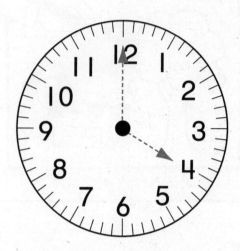

2.
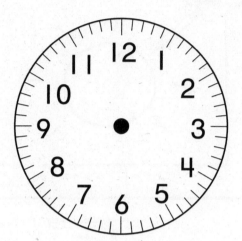

3.
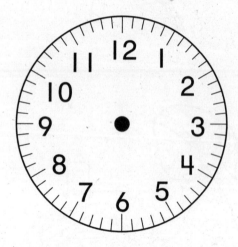

4.
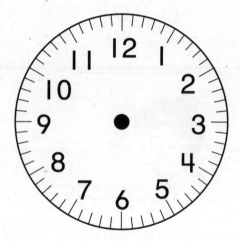

Draw hands on each clock to show the following times: 1) 4 o'clock, 2) 2 o'clock 3) 8 o'clock, 4) 11 o'clock.

Grade K, Chapter 9, Lesson 9.6

Problem-Solving Strategy: Use Logical Reasoning

The picture shows the last part of a story. Draw what happened first. Draw what happened next.

Name: _____

Mixed Review

Practice 58

_____

_____

_____

_____

_____

_____

Use numbers 1, 2, and 3 to order the events.

58 • Practice

Grade K, Chapter 10, Lesson 10.1

Length

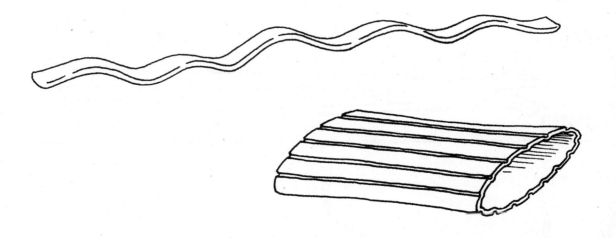

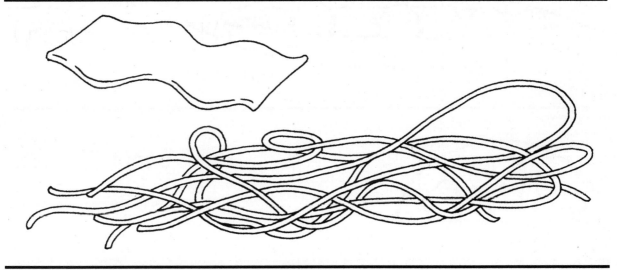

Color the object that is shorter. Draw a ring around the object that is longer.

Grade K, Chapter 10, Lesson 10.2

Measure Length

Practice 60

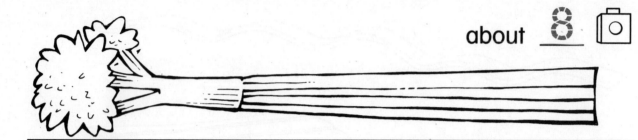 about 8 📷

 about ___ 📷

 about ___ 📷

 about ___ 📷

 about ___ 📷

Use 📷 to measure. Write about how many.

Practice 61

Weight

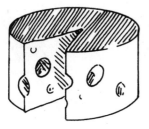	

	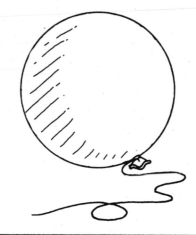

Draw a ring around the thing that is lighter than the first object in each row.

Name:

Capacity

Practice 62

 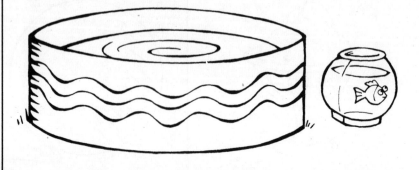

Use a blue crayon. Color the water in the container that holds less than the first container.

Name:

Temperature

Practice 63

Draw a ring around all hot objects. Draw an X on all cold objects.

Grade K, Chapter 10, Lesson 10.6

Name:

Explore Fair Shares

Practice 64

  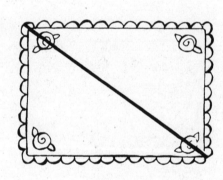

Find the shapes that show fair shares. Color one share.

Grade K, Chapter 10, Lesson 10.7

Name: _____

Practice 65

Halves of Regions

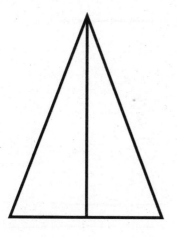 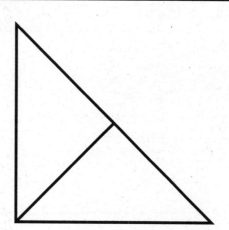 

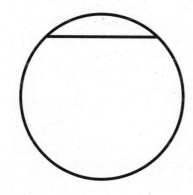 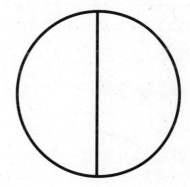 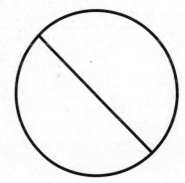

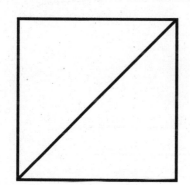 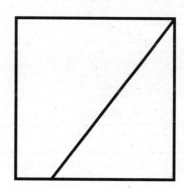 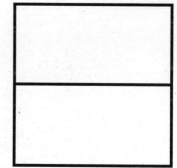

Draw a circle around each shape that shows halves.

Grade K, Chapter 10, Lesson 10.8

Problem-Solving Strategy: Guess and Test

Practice 66

Name:

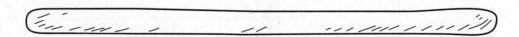

Draw a ring around the breadstick you guess is shorter. Measure with cubes to test your guess. Then color the breadstick that is longer.

Name:

Mixed Review

Practice 67

Color the animal that weighs more. Color the shorter part of the fence.
Color the taller tree. Color half of the ball.

Grade K, Chapter 11, Lesson 11.1

Find Two Parts

Practice 68

7

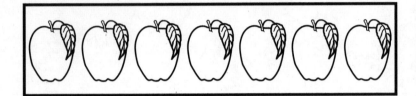 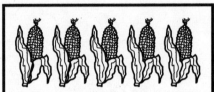

6

 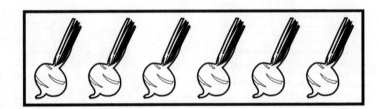

4

Choose the box that shows the number. Use two different colors to show two parts.

Grade K, Chapter 11, Lesson 11.2

Name:

Practice 69

Find a Total

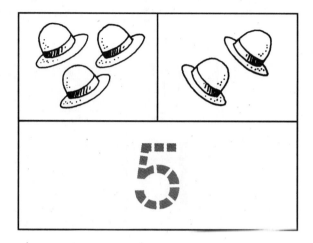
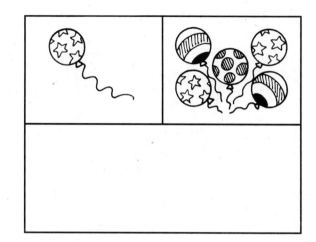

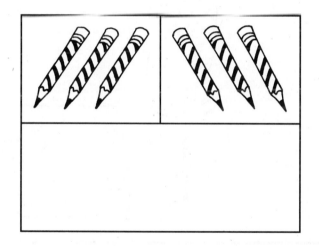

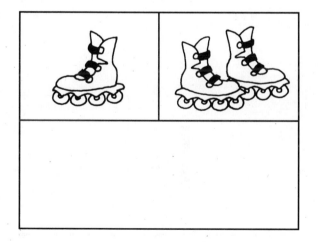
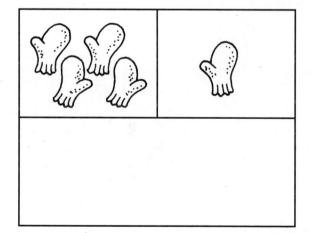

Write the total in each box. Draw a line to another box with the same total but different parts.

Grade K, Chapter 11, Lesson 11.3

Name: _____

Practice 70

Beginning to Add

2

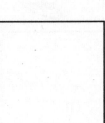
1 ☐

1

1 ☐

4
1 ☐

For the first two pictures, write the total in the box. For the third picture, draw to show two parts. Then write the total in the box.

70 • Practice

Grade K, Chapter 11, Lesson 11.4

Name: _____

Adding

Practice 71

$$4 + 2 = \underline{6}$$

$$6 + 1 = \underline{}$$

$$\underline{} + \underline{} = \underline{}$$

Draw the missing part. Add. Write the total. Then draw your own picture to show addition. Write an addition sentence to go with your picture.

Grade K, Chapter 11, Lesson 11.5

Name: _____

Beginning to Subtract

Practice 72

4

5

3

Cross out one or more than one. Write the part that is left in the box.

72 • Practice　　　　　　　　Grade K, Chapter 11, Lesson 11.6

Name: _____

Subtracting

Practice 73

3 − 1 = ___

5 − 1 = ___

___ − ___ = ___

Cross out one to subtract. Write the number that is left.
Draw a picture and cross out one to subtract. Write the subtraction sentence.

Grade K, Chapter 11, Lesson 11.7

Problem-Solving Strategy: Choose the Operation

4 __ 1 = 3

7 __ 2 = 9

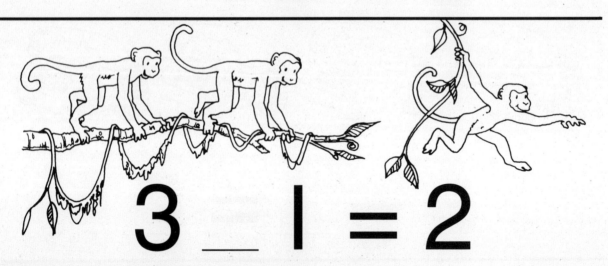

3 __ 1 = 2

Write a plus sign or a minus sign to finish each sentence.

Name:

Mixed Review

Practice 75

___ + ___ = ___

2 − 1 = ___

Write the addition sentence to go with the picture. Draw a picture to go with the subtraction sentence and write the answer.

Grade K, Chapter 12, Lesson 12.1

Practice • 75

Numbers 13 to 15

Name:

Practice 76

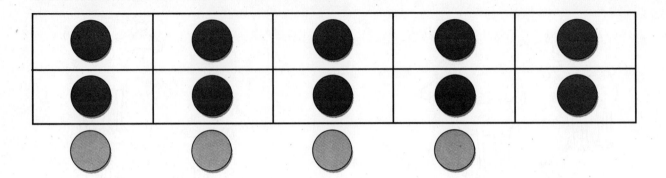

13   14   15

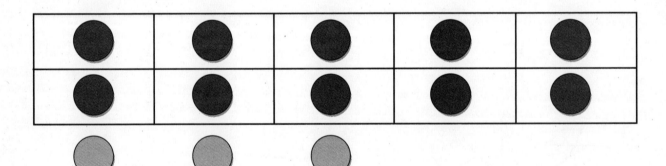

13   14   15

Draw a group to show how many. Draw a ring around the number.

76 • Practice

Grade K, Chapter 12, Lesson 12.2

Numbers 16 to 19

Draw stripes on each cat to show the number.

Name:

Numbers to 31

Practice 78

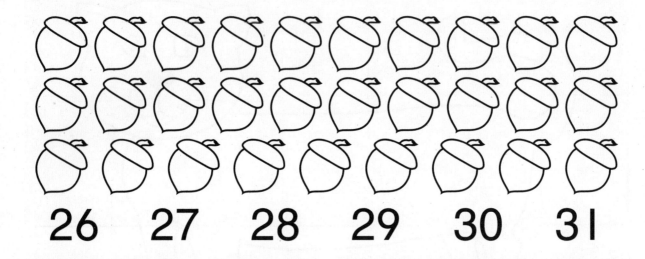

26 27 28 29 30 31

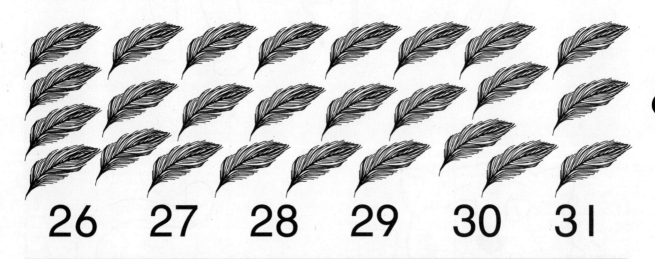

26 27 28 29 30 31

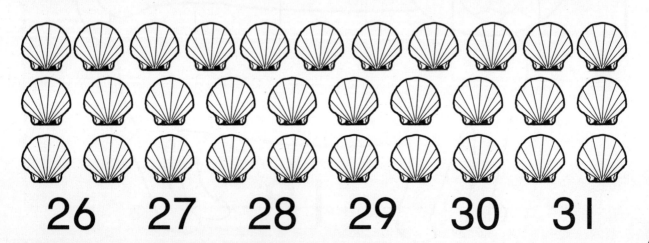

26 27 28 29 30 31

Draw a ring around the number that tells how many.

Name:

Calendar

Practice 79

Sunday	Monday	Tuesday	Wednesday	Thursday	Friday	Saturday
1	2	3				7
8			11	12	13	
15			18	19	20	21
22		24				
29		31				

Write the missing numbers.

Grade K, Chapter 12, Lesson 12.5

Problem-Solving Strategy: Draw a Picture

Practice 80

Name:

15 16 17 18 | 24 25 26 27

Draw a picture to solve each problem: 1) There are 7 scoops on one ice cream cone. There are 9 scoops on the other. How many scoops are there in all?
2) There are 12 feathers on one bird. There are 15 feathers on the other. How many feathers are there in all? Mark each number that tells how many in all.